ALTERNATOR BOOKS™

WEIRD ANIMALS

Brianna Kaiser

Lerner Publications ◆ Minneapolis

For everyone in my family, including our weird dog, little Jack

Lerner Publications Company
An imprint of Lerner Publishing Group, Inc.
241 First Avenue North
Minneapolis, MN 55401 USA

For reading levels and more information, look up this title at www.lernerbooks.com.

Main body text set in Aptifer Sans LT Pro.
Typeface provided by Linotype AG.

Editor: Lauren Foley **Designer:** Mary Ross **Photo Editor:** Annie Zheng

Library of Congress Cataloging-in-Publication Data

Names: Kaiser, Brianna, 1996– author.
Title: Weird animals / Brianna Kaiser.
Description: Minneapolis, MN : Lerner Publications, [2024] | Series: Wonderfully weird (Alternator books) | Includes bibliographical references and index. | Audience: Ages 8–12 | Audience: Grades 4–6 | Summary: "From funky looks to odd behaviors, catch an inside glimpse of some of the world's strangest animals. Young readers will have fun learning all about what makes these creatures unique"— Provided by publisher.
Identifiers: LCCN 2022033574 (print) | LCCN 2022033575 (ebook) | ISBN 9781728490700 (lib. bdg.) | ISBN 9798765601792 (eb pdf)
Subjects: LCSH: Animal behavior—Juvenile literature. | Animals—Miscellanea—Juvenile literature.
Classification: LCC QL751.5 .K333 2024 (print) | LCC QL751.5 (ebook) | DDC 591.5—dc23/eng/20220719

LC record available at https://lccn.loc.gov/2022033574
LC ebook record available at https://lccn.loc.gov/2022033575

Manufactured in the United States of America
1-53001-51019-9/30/2022

TABLE OF CONTENTS

INTRODUCTION:

FISH CAN . . . FLY?

YOU ARE ON A BOAT IN THE MIDDLE OF THE OCEAN WHEN YOU SEE SOMETHING SURPRISING. Fish are flying out of the water! They jump out of the water and into the air.

Flying fish have special pectoral fins. They don't actually fly. But they can swim at speeds of more than 35 miles (56 km) per hour to break the ocean's surface and launch into the air. Their pectoral fins allow them to glide in the air for up to 650 feet (198 m).

There are millions of animal species in the world. Some, like flying fish, can seem more unusual than other kinds of animals. They can be weird! Animals can be weird in many kinds of ways, including how they act and look.

CHAPTER 1:

ONE OF A KIND

ALTHOUGH THERE ARE MILLIONS OF ANIMALS, SOME STICK OUT AMONG OTHERS LIKE THEM! Whether because of how they look or how they grow, these animals stand out from the rest of their kind.

PANGOLIN

Pangolins are covered from head to tail in overlapping, plate-like scales. Because of their scales, many people confuse them for reptiles. But they are mammals—the only mammals covered in scales.

When pangolins feel threatened, they curl up into a tight ball. Then they use their tail, covered in sharp scales, to defend themselves. There are eight species of pangolins. Four of them live in Africa, and four live in Asia.

A pangolin rolling into a ball for safety

A Venomous Mammal

Some animals have venom. It can poison humans or other animals. Solenodons are one of a few kinds of mammals with venom. They inject venom through their teeth.

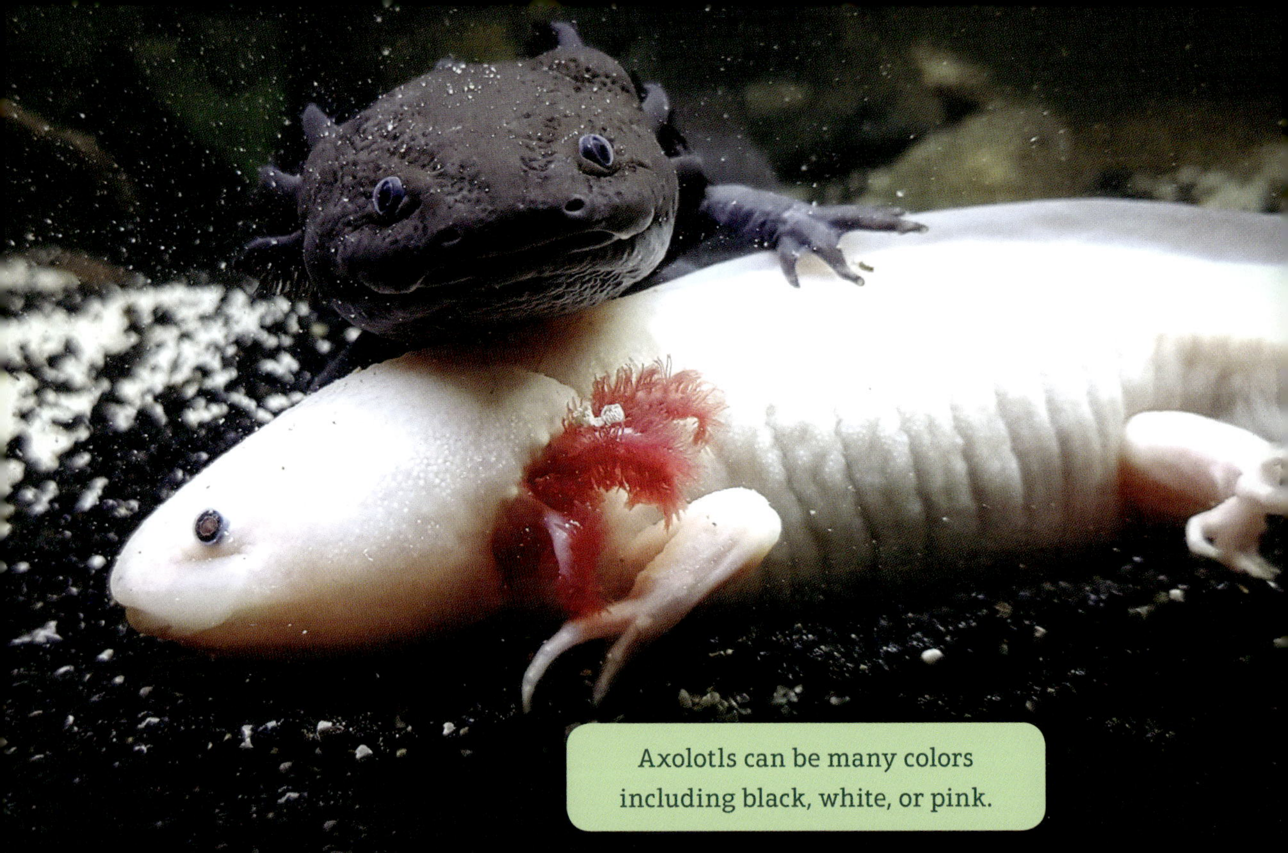

Axolotls can be many colors including black, white, or pink.

AXOLOTL

Salamanders and other kinds of amphibians usually go through metamorphosis. That is when many amphibians lose their gills and grow lungs because they are switching from living life in the water to living life on land. But one type of salamander, the axolotl, is different. Axolotls spend their lives in the water. And though they grow lungs too, they never lose their gills.

Keeping their gills isn't the only reason these animals are weird. Axolotls can regrow limbs and replace parts of their brains. They live in Lake Xochimilco in the Valley of Mexico and in the canals of Mexico City.

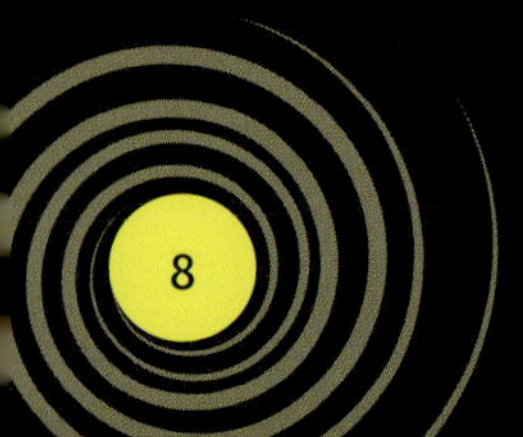

Japanese Spider Crab

There are around sixty thousand species of crustaceans, and the Japanese spider crab is the largest. From claw to claw, Japanese spider crabs are 12.5 feet (3.8 m). They weigh up to 44 pounds (20 kg).

They have ten legs but only walk on eight. The other two legs have chelipeds, or claws. Male Japanese spider crabs have longer chelipeds than walking legs, while the females' chelipeds are shorter than their walking legs. If any of these crabs loses a leg, they can regrow it.

Japanese spider crabs live on the Pacific seafloor near Japan. Some scientists believe these crabs can live fifty to one hundred years.

A Japanese spider crab walks underwater.

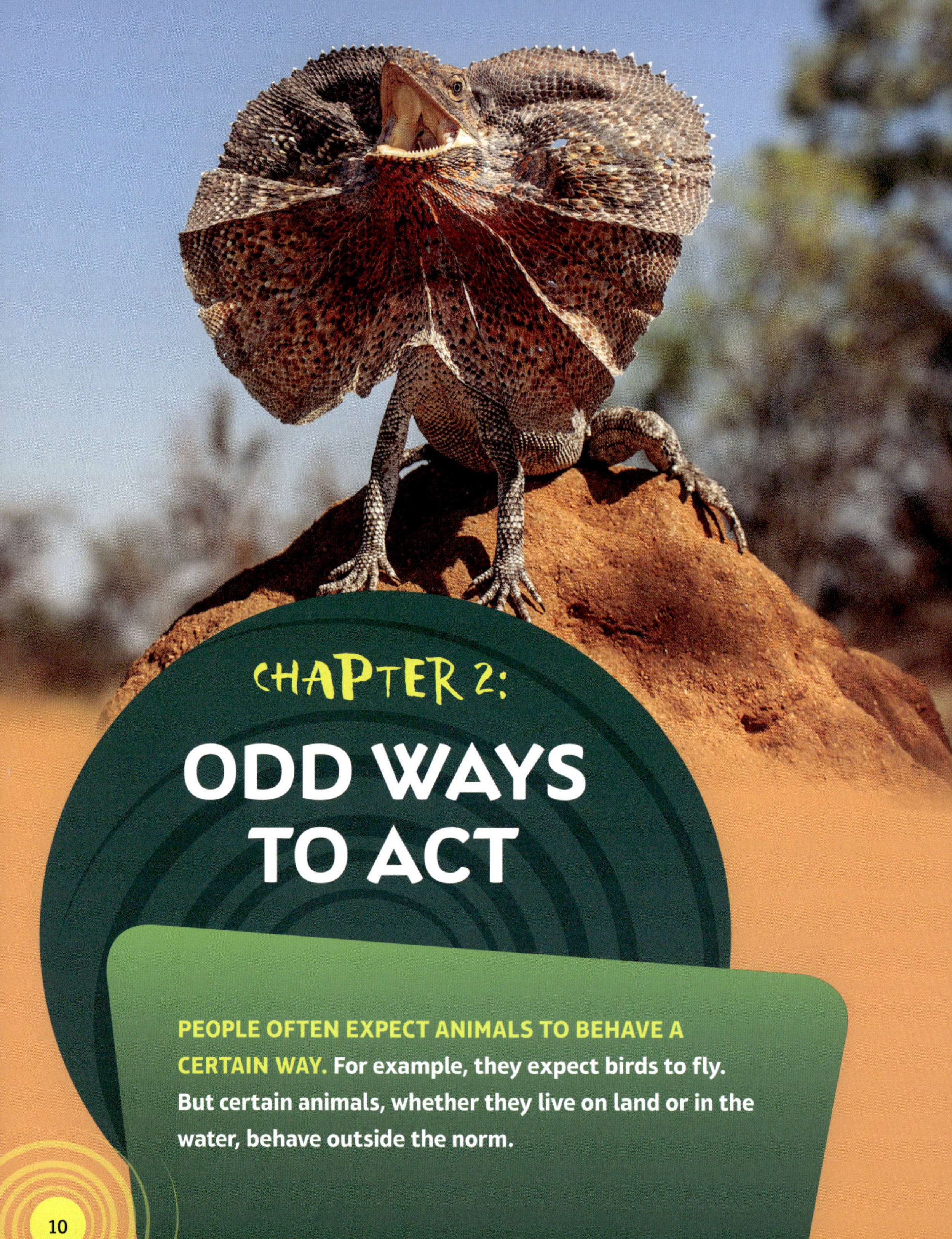

ODD WAYS TO ACT

PEOPLE OFTEN EXPECT ANIMALS TO BEHAVE A CERTAIN WAY. For example, they expect birds to fly. But certain animals, whether they live on land or in the water, behave outside the norm.

FRILLED LIZARD

Frilled lizards live in northern Australia. They are about 3 feet (0.9 m) long and weigh around 1 pound (0.4 kg). They spend a lot of time in trees.

What makes these lizards unique is how they defend themselves from predators. If a frilled lizard feels threatened, it will rise onto its hind, or back, legs. Then, as it opens its mouth to hiss, the lizard will open up a pleated skin flap that surrounds its head. Sometimes, the strange look still won't scare a predator. When that happens, the lizard will run on its hind legs to a nearby tree for safety.

SEA PIG

On the deep, muddy ocean floor, sea cucumbers use the tentacles around their mouth to search for food. But these aren't just any sea cucumbers. They are sea pigs. They were named for their pink coloring.

Sea pigs may have a weird look, but they also have a strange relationship with juvenile king crabs. Researchers have discovered that sea pigs allow juvenile king crabs to cling to them for rides. Researchers aren't certain why they do this. But they believe the crabs may cling to sea pigs for protection from predators.

Young king crabs often ride on the sides or bottom of the sea pig.

DISCOVERING NEW SPECIES

In 2022 scientists thought they might have found a new species of giant tortoise in the Galápagos Islands. To try to determine whether they found a new species or a population of an existing species, scientists compared blood samples and DNA from living and dead giant tortoises. The scientists were not able to confirm whether they found a new or existing species. But they were still excited to find more giant tortoises.

AFRICAN CLAWED FROG

Did you know that not all frogs hop? The African clawed frog has small front limbs with nonwebbed fingers and large hind legs that are webbed. These frogs are great swimmers, but they are not great hoppers. Instead, they crawl to get around on land.

African clawed frogs often have multicolored skin that includes greens, browns, and yellows. They can even change their coloring to blend into their background. This helps them hide from predators in their habitat. These frogs mainly live in eastern and southern Africa but can be found in freshwater areas around the world.

Hidden in Plain Sight

Some kinds of animals can easily blend with their surroundings. The satanic leaf-tailed gecko looks like a leaf. The great potoo looks like a tree stump.

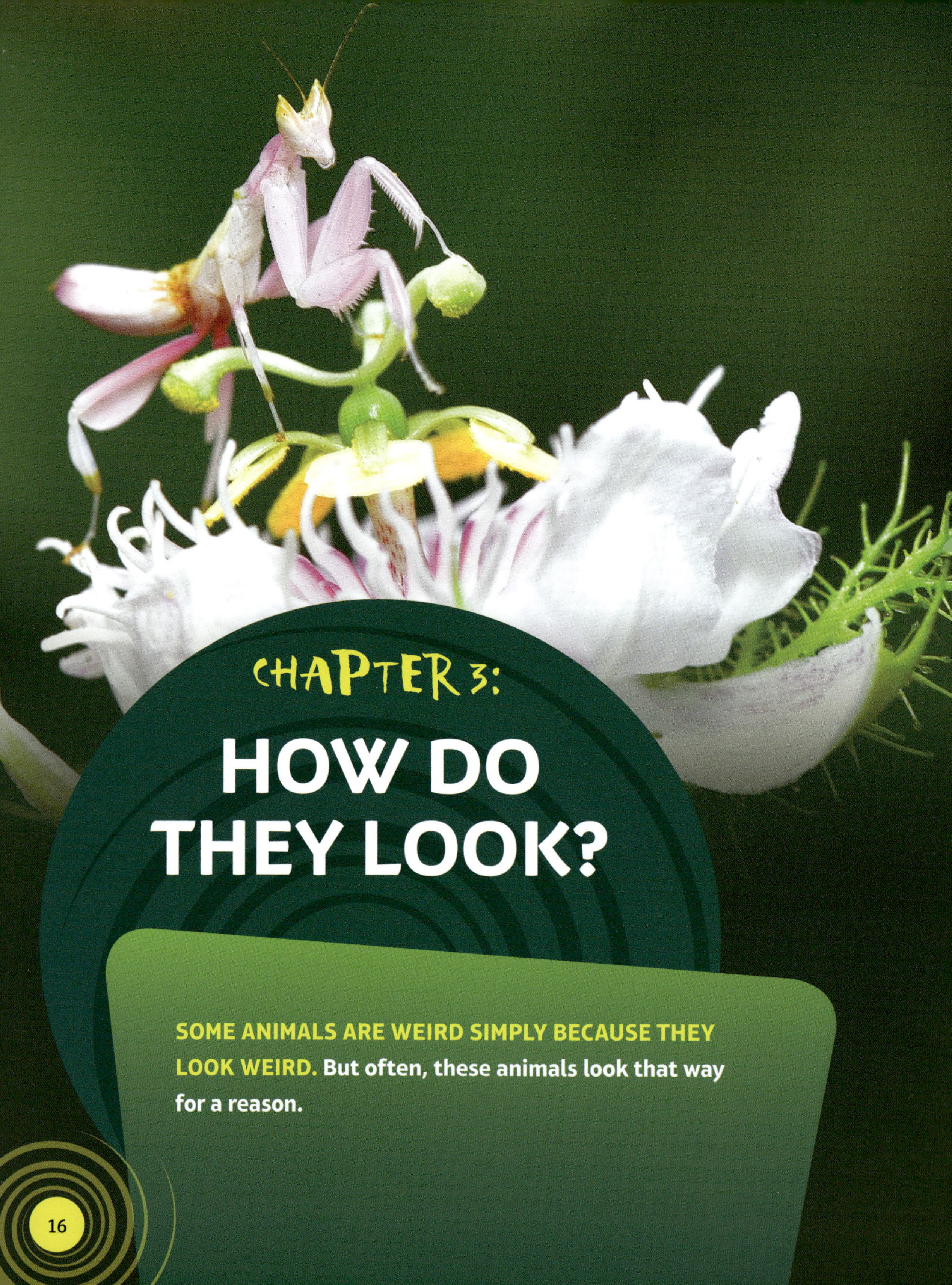

HOW DO THEY LOOK?

SOME ANIMALS ARE WEIRD SIMPLY BECAUSE THEY LOOK WEIRD. But often, these animals look that way for a reason.

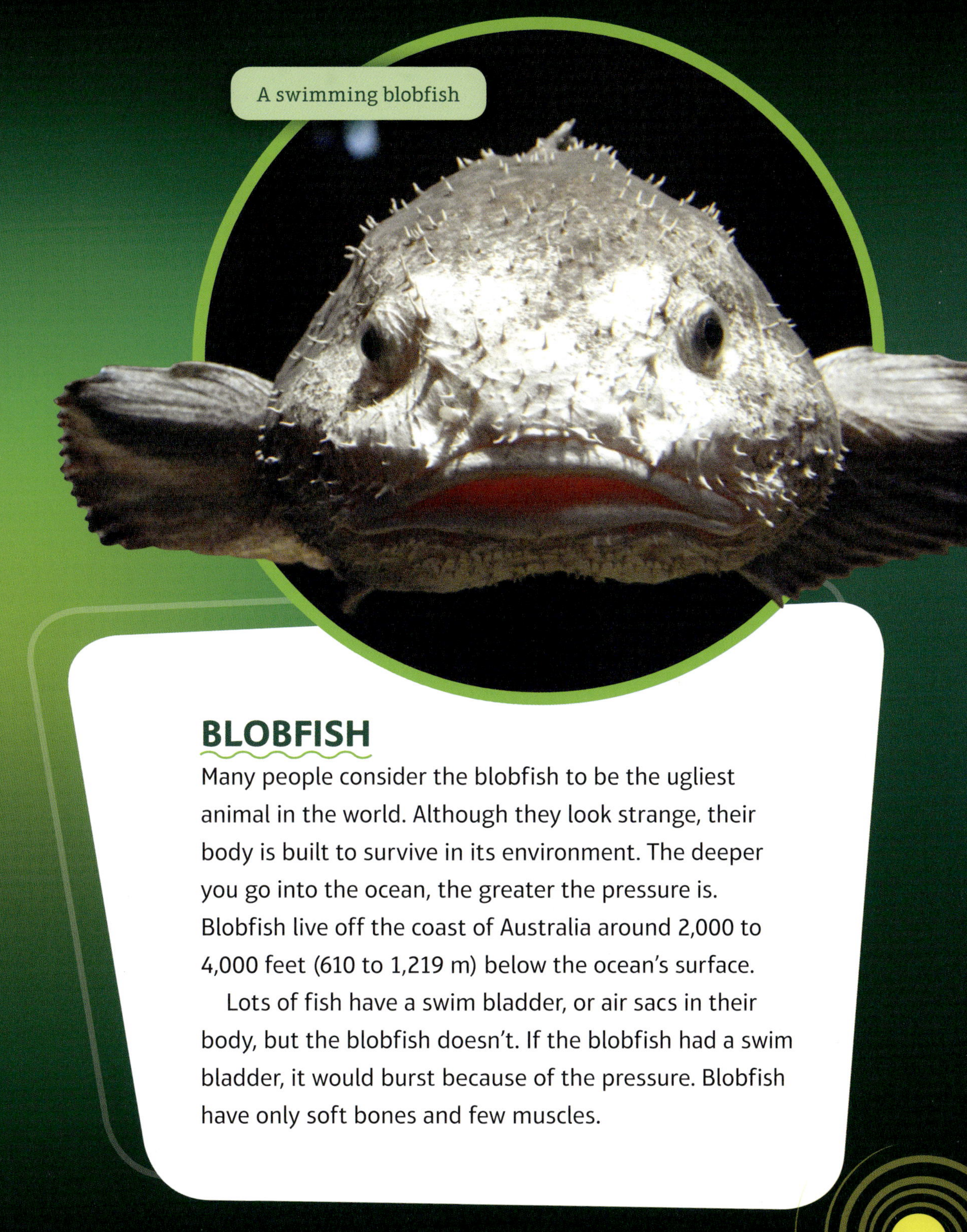

BLOBFISH

Many people consider the blobfish to be the ugliest animal in the world. Although they look strange, their body is built to survive in its environment. The deeper you go into the ocean, the greater the pressure is. Blobfish live off the coast of Australia around 2,000 to 4,000 feet (610 to 1,219 m) below the ocean's surface.

Lots of fish have a swim bladder, or air sacs in their body, but the blobfish doesn't. If the blobfish had a swim bladder, it would burst because of the pressure. Blobfish have only soft bones and few muscles.

RETICULATED GLASS FROG

Reticulated glass frogs are not made of glass, but they have see-through bellies. We can see some of their organs such as their heart. The spotted pattern on their backs is usually light green, yellow, or white. Scientists believe glass frogs' skin acts as camouflage to help them hide from predators.

They are very small, only about 1 inch (2.5 cm) long. They live in rain forests in Central America and South America.

A reticulated glass frog's organs are visible from below.

Pretend Flowers

The orchid mantis has similar coloring as orchid flowers, and its legs look like orchid petals. By looking like a flower, the mantis can lure in prey.

immortal Jellyfish

Many kinds of jellyfish live in the ocean. Some can glow in the dark. The Turritopsis dohrnii is one of them. But the longer you look at them, the stranger these jellyfish are.

These jellyfish are also known as the immortal jellyfish. That is because they are able to change their adult cells into child cells. Although the immortal jellyfish has the ability to live forever, most of them don't. They can still die from many causes such as injury or attacks by predators.

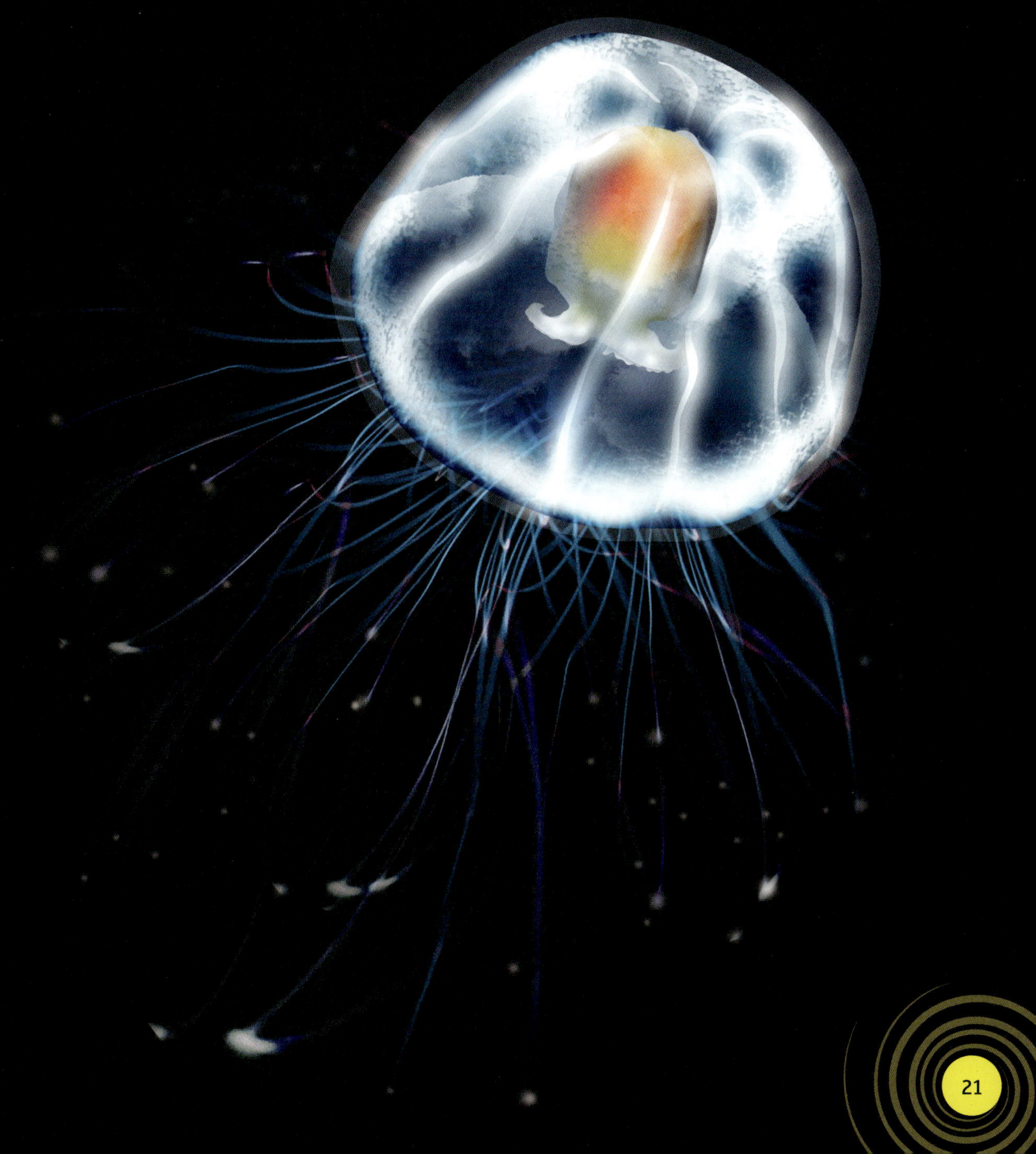

The immortal jellyfish might change its adult cells
to child cells if it feels threatened.

CHAPTER 4:
DANCING FOR MATES
A MATING DISPLAY IS WHEN AN ANIMAL, USUALLY A MALE, ACTS A CERTAIN WAY TO ATTRACT A MATE. Some animals have weirder mating displays than others.

SUPERB BIRD OF PARADISE

The superb bird of paradise is black with blue on its neck. But the males greatly change their appearance when they perform a mating dance. They do the mating dance to attract a female mate. During the dance, the male fans out its feathers to form an oval shape. That makes its bright blue feathers stand out. Then it hops around the female.

In 2016 a researcher and a photographer found another superb bird of paradise species in New Guinea. These birds had a slightly different mating dance. The males fanned out their feathers to form a crescent shape instead of an oval, and they slid from side to side around the female instead of hopping.

A superb bird of paradise showing off its blue feathers to attract a mate

MAGNIFICENT FRIGATE BIRD

Magnificent frigate birds are black birds that can have a wingspan greater than 7 feet (2.1 m) long. They live on islands and ocean coasts. The magnificent frigate bird doesn't appear to be more unusual than any other bird. But that changes when a male is trying to attract a female.

While females have white throats and bellies, males have a bright red throat pouch. To attract a female, a male frigate bird will inflate its red throat pouch and drum on it with its bill.

A magnificent frigate bird puffs up its red throat pouch.

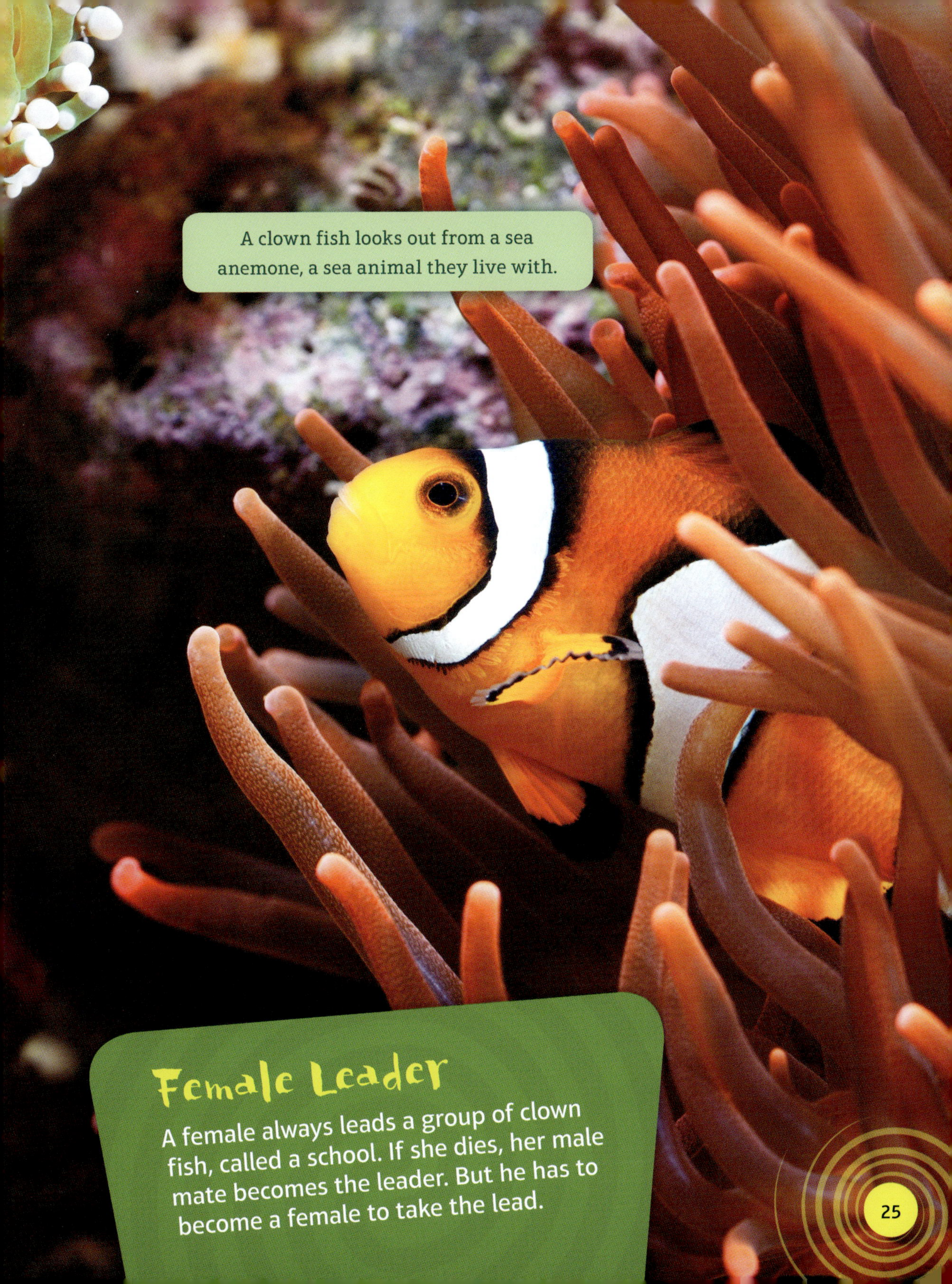

Female Leader

A female always leads a group of clown fish, called a school. If she dies, her male mate becomes the leader. But he has to become a female to take the lead.

PEACOCK SPIDER

There are more than forty-five thousand spider species. They live all over the world. Peacock spiders live mainly in Australia. Males have colorful bodies, but all the spiders are so small that it is hard to see all the colors without a big lens.

Peacock spiders have a special mating dance. The male spider will wave a part of its body, usually including its back end and a couple of legs, in front of a female. About ninety species of peacock spiders have been discovered. Each species' mating dance varies slightly.

A peacock spider waving its back legs and showing its colorful back

IMPACT OF CLIMATE CHANGE

Earth is warming. This negatively impacts animals in many ways. Many animals will have to leave their homes because their food sources are running out or their homes are disappearing. Others will have to adapt to their surroundings. Many animal species are becoming endangered or extinct because of climate change. People can help by learning more about climate change and raising awareness of the issue.

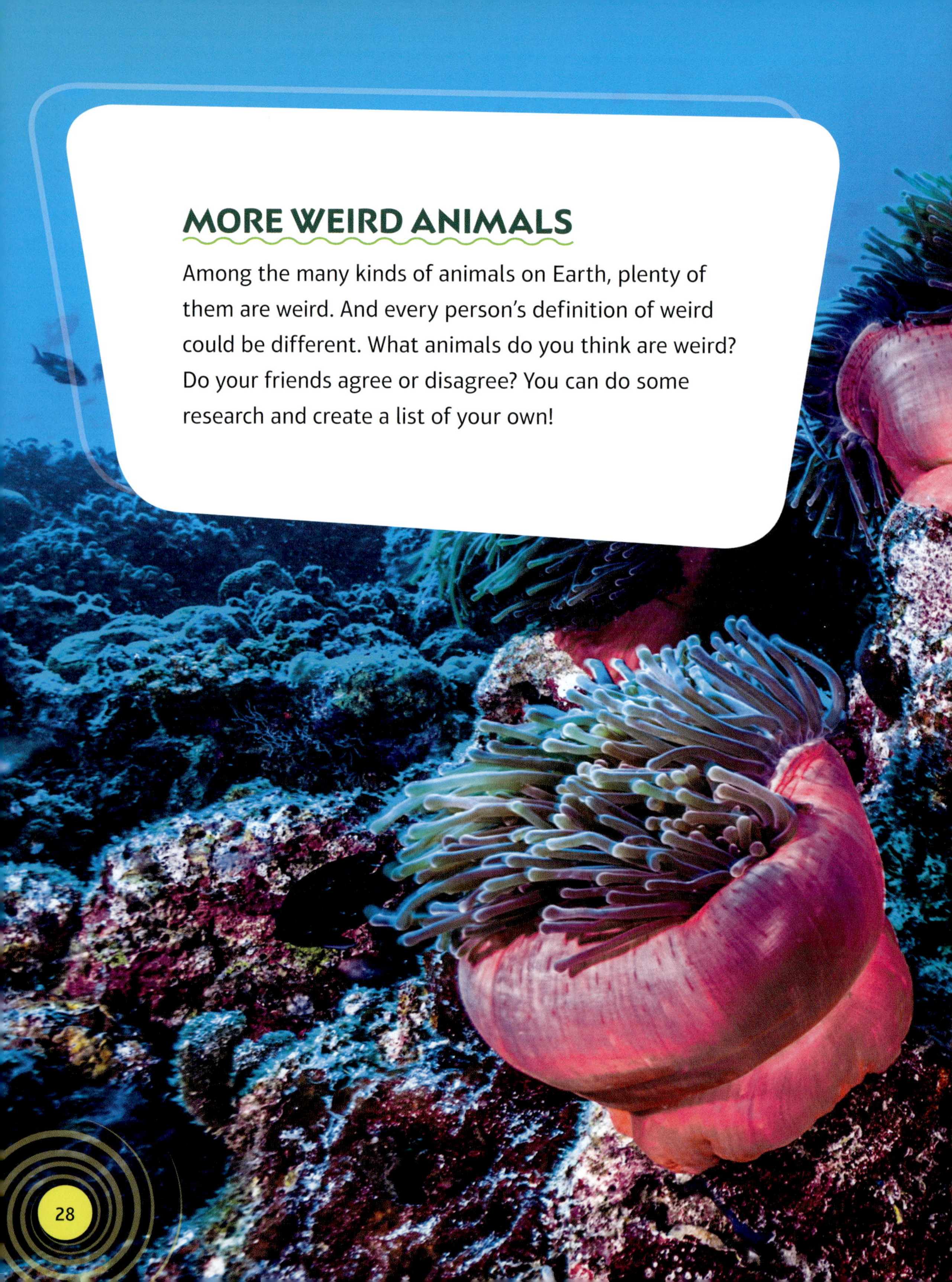

MORE WEIRD ANIMALS

Among the many kinds of animals on Earth, plenty of them are weird. And every person's definition of weird could be different. What animals do you think are weird? Do your friends agree or disagree? You can do some research and create a list of your own!

An underwater habitat. All kinds of habitats might have weird animals in them.

GLOSSARY

amphibian: a small animal that often spends part of its life in water and part of its life on land

camouflage: using color as a way of hiding or blending into surroundings

immortal: living forever

mammal: a warm-blooded animal with a skeleton and fur or hair on its skin

mate: a partner for an animal

metamorphosis: the act of a living thing changing as it grows

predator: an animal that hunts other animals for food

pressure: a strong force

prey: an animal that is hunted as food by another animal

species: a group of living things that have similar characteristics

LEARN MORE

Animal Facts for Kids
https://kids.kiddle.co/Animal

Bassier, Emma. *Japanese Spider Crabs*. North Mankato, MN: DiscoverRoo, 2020.

Eszterhas, Suzi. *Operation Pangolin: Saving the World's Only Scaled Mammal*. Minneapolis: Millbrook Press, 2023.

Hyde, Natalie. *Animal Oddballs*. New York: Crabtree, 2020.

National Geographic Kids: The Ocean's Weirdest Creatures!
https://www.natgeokids.com/uk/discover/animals/sea-life/strange-sea-creatures/

National Geographic Kids: Really Weird Animals
https://kids.nationalgeographic.com/videos/topic/really-weird-animals

Twiddy, Robin. *Silly Animal Science*. Minneapolis: Lerner Publications, 2022.

Weird Animals
https://easyscienceforkids.com/weird-animals/

INDEX

PHOTO ACKNOWLEDGMENTS

Image credits: blickwinkel/Alamy Stock Photo, p. 4; Artem Avetisyan/Shutterstock, p. 6; Imagevixen/RooM/Getty Images, p. 7; Jeffrey Lagmay/Shutterstock, p. 8; Jakkapan Prammanasik/Moment Open/Getty Images, p. 9; Ken Griffiths/iStock/Getty Images, p. 10; Jean-Paul Ferrero/AUSCAPE/Alamy Stock Photo, p. 11; NOAA/MBARI/Wikimedia Commons (CC BY-SA 3.0), p. 12; Milan Zygmunt/Shutterstock, p. 14; Ignacio Palacios/Stone/Getty Images, p. 15; Eko Budi Utomo/Shutterstock, p. 16; AP Photo/Kyodo, p. 17; Allen Lara Gonzalez/Shutterstock, p. 18; YoONSpY/Shutterstock, p. 19; scubadesign/Shutterstock, p. 20; Adisha Pramod/Alamy Stock Photo, p. 21; chonlasub woravichan/Shutterstock, p. 22; Corbin17/Alamy Stock Photo, p. 23; Agami Photo Agency/Shutterstock, p. 24; Alex Stemmers/Shutterstock, p. 25; crbellette/Shutterstock, p. 26; Giordano Cipriani/The Image Bank/Getty Images, p. 28.

Design elements: amgun/Shutterstock.

Cover image: Ken Griffiths/Shutterstock.